Disclaimer

The publisher of this book is by no way associated with the National Institute of Standards and Technology (NIST). The NIST did not publish this book. It was published by 50 page publications under the public domain license.

50 Page Publications.

Book Title: Tracking the National Fire Problem: The Data behind the Statistics

Book Author: Douglas S. Thomas; David T. Butry;

Book Abstract: The objective of this report is to support the development of descriptive statistics and associated measures of uncertainty for characterizing, tracking, and better understanding the root causes of the U.S. fire burden by identifying the relevant costs and losses associated with fire,i.e., information needed to make cost-effective decisions for reducing the economic impact fires have on society,and identifying sources of data that would allow future statistical analysis, while highlighting areas where future research is needed. This data will be used to develop performance metrics, enabling future comparisons between the use of new fire mitigation technologies and their impact on the U.S. fire burden.

Citation: NIST TN - 1717

Keywords: fire burden; fire statistics; fire data; mitigation costs; fire cost; fire loss; economics; cost plus loss

U.S. Department of Commerce
National Institute of Standards and Technology

Applied Economics Office
Engineering Laboratory
Gaithersburg, MD 20899

Tracking the National Fire Problem:
The Data Behind the Statistics

Douglas S. Thomas
David T. Butry

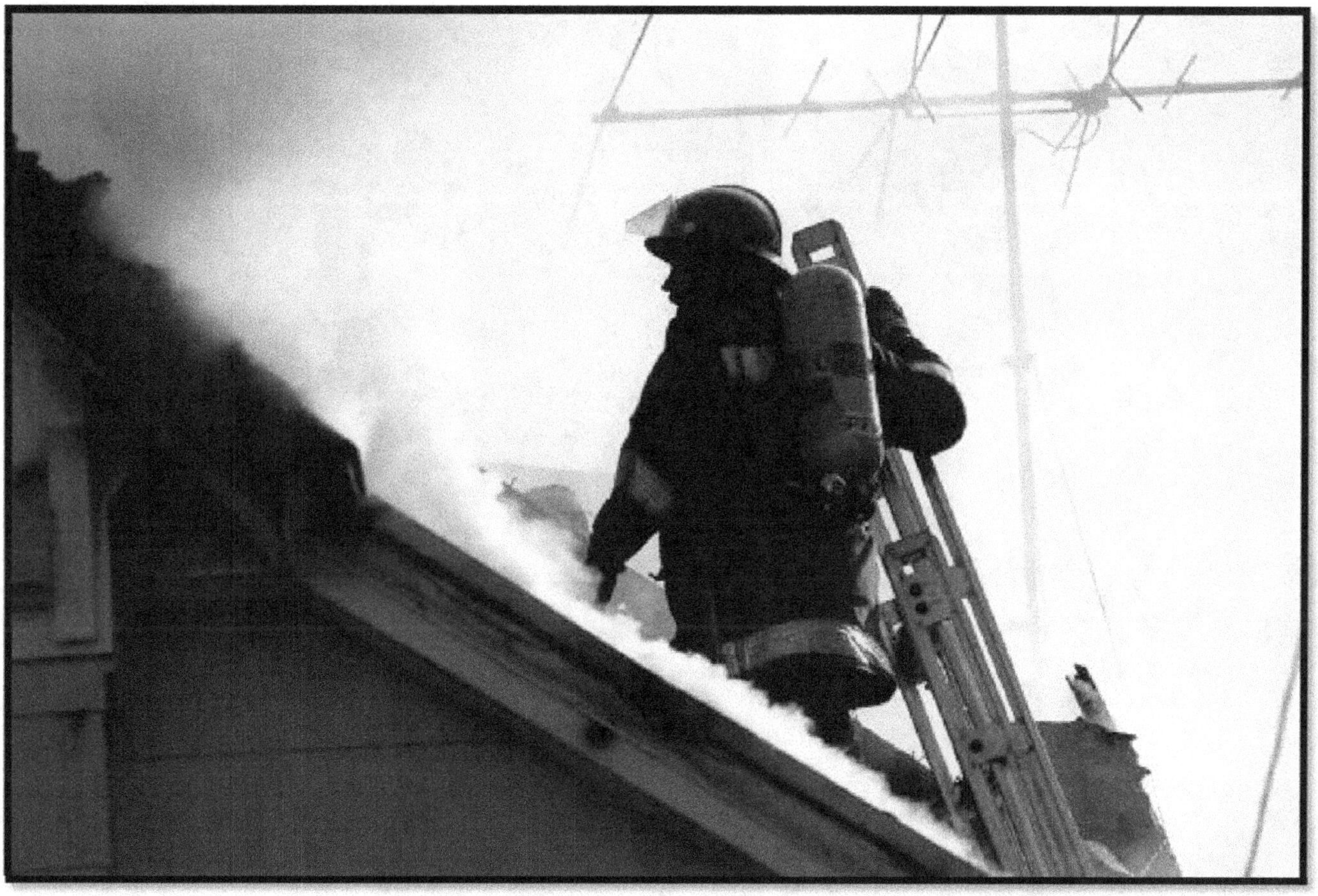

U.S. Department of Commerce
National Institute of Standards and Technology

Applied Economics Office
Engineering Laboratory
Gaithersburg, MD 20899

Tracking the National Fire Problem:
The Data Behind the Statistics

Douglas S. Thomas
David T. Butry

Sponsored by:
National Institute of Standards and Technology
Engineering Laboratory

September 2011

U.S. DEPARTMENT OF COMMERCE
Rebecca M. Blank, Acting Secretary

NATIONAL INSTITUTE OF STANDARDS AND TECHNOLOGY
Patrick D. Gallagher, Under Secretary of Commerce for Standards and Technology and Director

Abstract

The objective of this report is to support the development of descriptive statistics and associated measures of uncertainty for characterizing, tracking, and better understanding the root causes of the U.S. fire burden by identifying the relevant costs and losses associated with fire—i.e., information needed to make cost-effective decisions for reducing the economic impact fires have on society—and identifying sources of data that would allow future statistical analysis, while highlighting areas where future research is needed. These data will be used to develop performance metrics, enabling future comparisons between the use of new fire mitigation technologies and their impact on the U.S. fire burden.

Keywords: cost plus loss, economics, fire burden, fire cost, fire data, fire loss, fire statistics, mitigation costs

Preface

This study was conducted by the Applied Economics Office in the Engineering Laboratory at the National Institute of Standards and Technology. The study provides a synopsis of available data depicting the U.S. fire burden.

Disclaimer

Certain trade names and company products are mentioned in the text in order to adequately specify the technical procedures and equipment used. In no case does such identification imply recommendation or endorsement by the National Institute of Standards and Technology, nor does it imply that the products are necessarily the best available for the purpose.

Cover Photographs Credits

Microsoft Clip Art Gallery Images used in compliance with Microsoft Corporation's non-commercial use policy.

Acknowledgements

The authors wish to thank all those who contributed so many excellent ideas and suggestions for this report. Special appreciation is extended to Stanley Gilbert and Jason Averill for their thorough reviews and many insights and to Carmen Pardo for her assistance in preparing the manuscript for review and publication. The authors also wish to thank Robert Chapman and Nicos S. Martys for their reviews.

TABLE OF CONTENTS

LIST OF FIGURES

LIST OF TABLES

1 Introduction

The objective of this report is to support the development of descriptive statistics and associated measures of uncertainty for characterizing, tracking, and better understanding the root causes of the U.S. fire burden (i.e., the costs and losses from unwanted fires) by identifying the relevant costs and losses associated with fire—i.e., information needed to make cost-effective decisions for reducing the economic impact fires have on society— and identifying sources of data, that would allow future statistical analysis, while highlighting areas where future research is needed. These data will be used to develop performance metrics, enabling comparisons between the use of new fire mitigation technologies and their impact on the U.S. fire burden in a subsequent report from the National Institute of Standards and Technology (NIST). In some cases techniques are used to estimate costs or losses; since this report focuses on identifying sources of data, these techniques are discussed only briefly.

Advancements in measurement science are needed to accurately measure, characterize the uncertainty, and track the U.S. fire burden. Currently, detailed annual statistics characterizing fire burden are not published in a way to support economic analysis of the entire U.S. fire problem, nor are associated measures of uncertainty provided. Without improved accuracy and estimates of uncertainty, it is difficult to assess year-to-year variations (e.g., trends) in key fire statistics, and to evaluate the impact new technologies or fire mitigation strategies have on fire losses, injuries, and fatalities, and also to assess their cost-effectiveness.

The U.S. Fire Administration (USFA), National Fire Protection Association (NFPA), and the Consumer Products Safety Commission (CPSC) all provide important reports on the fire problem in the United States. The NFPA and CPSC administer surveys to collect data to augment that provided in NFIRS (USFA) or to collect data not collected by NFIRS. Because NFIRS is a voluntary (i.e., non-random) data collection system, however, the statistics derived from it may not be representative of the U.S. fire problem. Thus, statistics derived from NFIRS data may not be comparable from year-to-year, making the tracking of fire trends difficult. Another challenge is that there is not a method for measuring, or analyzing, the uncertainty (i.e., measurement error) surrounding current fire statistics. These are difficult to create given the non-randomness of the NFIRS sample.

The remainder of the report is organized as follows: Section 2 presents an economic model of optimal fire protection, which provides the motivation for collecting and analyzing fire-related cost and loss data; Section 3 catalogues the major individual components of cost and loss in the U.S., and identifies the stakeholder groups most affected, thereby highlighting how individual costs and losses are distributed across society; Section 4 discusses the availability of cost and loss data and the kinds of data currently recorded; Section 5 focuses on the remaining data needs—those costs and losses not currently made available or those that are difficult to measure. Section 5 serves to motivate future data collection research on this topic.

2 Economic Model of Optimal Fire Protection

From an economic standpoint, the optimal investment level into goods and services that provide fire protection and mitigation is that which minimizes combined costs of fire protection and mitigation plus the resulting losses from fires that continue to occur. At the optimum, an additional unit of fire protection or mitigation will cost more than the economic value of the fire avoidance.

The cost plus loss minimization objective function can be written as:

$$Min\ H = C + L,$$

where C are the total costs to mitigate (or respond to) fire, and L are total economic value of losses occurring from fire.

Costs can be decomposed into the following:

$$C = \sum_{i=1}^{N} P_i X_i,$$

where X is the quantity of fire risk reduction effort (techniques and technologies) i used to mitigate fire (e.g., one hour of firefighter labor, a smoke alarm), P is the unit price of risk reduction effort i (e.g., hourly dollar equivalent of one hour of firefighter labor, cost per smoke alarm), and there is a total of N risk reduction techniques and technologies (effort) available.

Losses can be decomposed into the following:

$$L = \sum_{j=1}^{J} W_j(X,Z)D_j(X,Z),$$

where D is the physical quantity of loss (damage) resulting from fire j, W is the economic value of loss (per unit D) resulting from fire j, and Z is the set of exogenous factors that affect the physical quantity of loss and the economic value of loss.[1] The physical quantity of loss and its economic value are functions of the quantity of fire risk reduction effort and other exogenous factors (e.g., time of fire, weather). It's assumed that the marginal effect of any one fire risk reduction technique or technology on the physical and economic value of damage is non-positive (i.e., fire mitigation effort reduces the number and severity of fires or is ineffective):

$$\frac{\partial W_j}{\partial X_i} \leq 0, \qquad \forall i, j$$

and

$$\frac{\partial D_j}{\partial X_i} \leq 0, \qquad \forall i, j.$$

Thus, the cost plus loss minimization objective function can be re-written as:

$$\min_{X} H = \sum_{i=1}^{N} P_i X_i + \sum_{j=1}^{J} W_j(X, Z) D_j(X, Z).$$

Minimizing the objective function, H, with respect to fire risk reduction effort, X_i, yields the following N first-order conditions:[2]

---

[1] Z contains factors that affect both W and D, only W, and only D.
[2] While not shown, for a minimum to occur the second-order conditions basically require that the quantity of risk reduction effort (X) must exhibit diminishing marginal returns with respect to reducing fire losses (W) and fire damages (D).

$$\frac{\partial H}{\partial X_1} = P_1 + \sum_{j=1}^{J} \frac{\partial W_j}{\partial X_1} D_j + W_j \frac{\partial D_j}{\partial X_1} = 0$$

$$\vdots$$

$$\frac{\partial H}{\partial X_N} = P_N + \sum_{j=1}^{J} \frac{\partial W_j}{\partial X_N} D_j + W_j \frac{\partial D_j}{\partial X_N} = 0$$

Setting the first-order conditions equal and solving for the total optimal physical damage level, D^* yields,[3]

$$D^* = \sum_{j=1}^{J} D_j^* = -\frac{\left(\sum_{i=i}^{N} P_i\right) + \left(\sum_{i=1}^{N} \frac{\partial D}{\partial X_i}\right) \sum_{j=1}^{J} W_j}{\left(\sum_{i=1}^{N} \frac{\partial W}{\partial X_i}\right)},$$

where to simplify the notation, the marginal effect effort i has on the physical damage of fire j is assumed constant across all J-fires, or $\frac{\partial W_1}{\partial X_i} = \frac{\partial W_2}{\partial X_i} = \cdots = \frac{\partial W_J}{\partial X_i} = \frac{\partial W}{\partial X_i}$ for every i, and the marginal effect effort i has on economic value of damage of fire j is assumed constant across all J-fires, or $\frac{\partial D_1}{\partial X_i} = \frac{\partial D_2}{\partial X_i} = \cdots = \frac{\partial D_J}{\partial X_i} = \frac{\partial D}{\partial X_i}$ for every i. This demonstrates the optimal level of physical fire damage is a function of the prices of risk reduction effort, the economic (unit) value of fire loss, and the marginal effects of risk reduction effort on physical fire damage and the economic value of fire damage. Such a model could be used to evaluate how changes in the effectiveness of firefighting technologies alter the societal optimal level of physical fire damage (or total economic value of fire damage if D^* is multiplied by a measure of P).

Solving the first-order conditions for the optimal level of investment into fire protection and mitigation is more difficult to demonstrate without additional insights into the

---

[3] This expression is a necessary, but not a sufficient condition of the minimization—i.e., there may be values of X that maintain the equality, but do not achieve the minimum. However, for those values of X that achieve the minimum, the equality must hold.

functional form specifications of $W(X, Z)$ and $D(X, Z)$. In addition, such information would be needed to numerically solve the above (theoretical) solution.

Parameterizing the functional forms of $W(X, Z)$ and $D(X, Z)$ could be accomplished using statistical methods; however, the success of such modeling efforts are predicated on the availability of observational data related to the economic value and physical losses associated with fire, information regarding the use and intensity of fire risk reduction techniques and technologies (effort), as well as data on other exogenous factors that affect fire-related losses—*i.e., needed are good cost and loss information on individual fires*. Without accurate and consistently measured data describing details of the U.S. fire problem, economically-optimal risk reduction strategies cannot be assessed, nor can the returns to investments into fire prevention and mitigation.

The focus of the remainder of the report is on cataloguing the more significant costs and losses related to fire, and identifying known data source that provides detail into their specifics, which might be used in future economic analysis.

3   Costs, Losses, and Stakeholders

In the U.S., fire has destroyed homes, neighborhoods, and communities along with vast amounts of grass and forest land. The nation has invested a significant amount of resources in preventing the destruction caused by fire. Local, state, and federal governments across the nation employ fire fighters and enlist volunteers to protect communities from fire. They have further mandated that structures be designed to meet standards that reduce the damage caused by fires. The threat that fire poses to humans and their environment necessitates the need to quantify the number and extent of fires in the U.S. along with the costs of prevention, response, and damage related to fire incidents. Individual stakeholders are affected differently by fire incidents; therefore, it is important to identify the types of costs and losses that can be incurred along with the stakeholders that bear them. The identification of these items can then direct the types of data needed to quantify costs and losses.

3.1   Costs and Losses

It is important to note the difference between costs and losses as these terms are often used interchangeably. For the purpose of this document, costs include expenditures to mitigate fire or respond to it whereas losses are the values of assets that are destroyed or damaged by fire. The types of costs that might be incurred include the costs of fire protection for constructed facilities or the investigation of a fire incident; these costs are categorized in Table 3.1. There are many costs that are not readily apparent to the casual observer (e.g., fire protection research and design), but these costs can be significant. Structures are often designed to reduce the impact of a fire; however, the means for this fire protection must be identified and tested before it is implemented into the general building stock. This also applies to equipment and goods sold to the general public. There is also the cost of training fire service personnel and emergency responders. Citizens often see fire fighters in action; however, they do not see the constant training that they engage in prior to responding to a fire. These costs are not seen by the average citizen; however, they often benefit from these expenditures.

7

Table 3.1: Categories of Costs and Losses Resulting from Fire Incidents

Costs	Losses
Fire protection research and design for constructed facilities	Direct property damage (net insurance)
Installation of fire protection in newly constructed facilities	Indirect property damage (net insurance)
Retrofitting constructed facilities for fire protection	Economic value of civilian fatalities
Maintenance and repair of facility fire protection	Economic value of civilian injuries
Fire protection research and design for equipment and goods	Economic value of fire fighter fatalities
Production of fire protection in equipment and goods	Economic value of fire fighter injuries
Fire service personel and training	Environmental degradation of the atmosphere
Fire service equipment	Degradation of the landscape
Fire protection of structures against WUI fires	Loss of recreational use
Evacuation services and planning	Timber loss
Emergency responder personel and training	Direct business interruption
Emergency responder equipment	Indirect business interruption
Wildland fuels management	Deadweight losses
Standards and codes	Psychological losses (irreplaceable goods)
Insurance premium	Transportation degradation
Fire prevention education/campaigns	Infrastructure losses
Aesthetic losses for fire protection design	Utility losses
Displacement costs (hotel and temporary shelter)	Wildlife losses
Fire code enforcement	Income Losses
Fire incident investigation	

In addition to unseen costs, there are also many losses that are not seen by the general public; deadweight losses and environmental degradation, for example. As seen in Table 3.1, some losses are direct losses; such as direct business interruption. This is the business lost due directly to the destruction or damage caused to a business. There are also indirect losses such as indirect business interruption, which includes the decline in business due to other's property being damaged. These indirect losses are often overlooked and not included when discussing the impact of a fire incident.

Summary of Costs

Fire protection research and design for constructed facilities: Expenditures on research and development have identified materials and structural designs that reduce the impact of a structural fire. These designs are often integrated into building codes and standards.

Installation of fire protection in newly constructed facilities: The incorporation of fire protection, such as fire codes and standards, in newly constructed facilities adds additional costs for materials and construction.

Retro-fitting constructed facilities for fire protection: Older structures often do not meet modern fire codes and standards. Occasionally, these facilities must be retro-fitted to meet modern codes.

Maintenance and repair of facility fire protection: There are costs associated with the maintenance of fire protection and suppression equipment installed in constructed facilities. These costs include system maintenance, testing, industrial fire brigades, and training programs for occupational fire protection and safety.

Fire protection research and design for equipment and goods: Many products are designed to reduce the likelihood or impact of a fire. Cigarettes, for example, are designed as Fire Safe Cigarettes (FSC) that extinguish more quickly than standard cigarettes reducing the likelihood of igniting other objects.

Production of fire protection in equipment and goods: Products that are designed to reduce the likelihood or impact of a fire often have increased costs of production. Automobiles for instance, have a fire wall protecting the occupants of the car from heat and/or fire in the engine compartment, which incurs costs for materials and installation.

Fire service personnel and training: In the event that a fire does occur, fire service personnel respond to these incidents and extinguish them. A significant amount of time and resources are expended training these individuals and maintaining specified levels of physical fitness and operational readiness.

Fire service equipment: Expenditures on protective equipment for fire service personnel as well as the equipment used to fight fires.

Fire protection of structures against wildland-urban interface (WUI) fires: Each year, wildland fires threaten numerous structures and occupants in close proximity to wildland areas. A significant amount of time and resources are dedicated to taking preventive measures to reduce the probability that a wildland fire will ignite a constructed facility.

Evacuation services and planning: Often times, communities and/or structures must be evacuated to ensure the safety of their citizens and/or occupants. A significant amount of time and resources are dedicated to planning, coordinating, and practicing these evacuations so that they can be executed rapidly and without incident.

Emergency responder personnel and training: Fires often injure civilians and fire fighters. These injuries often require the attention of emergency responder personnel. A significant amount of time and resources are expended training these individuals and maintaining certifications.

Emergency responder equipment: Injuries caused by fire often require the attention of emergency responder personnel from entities other than the fire service. These services require vehicles and equipment.

Wildland Fuels Management: Wildland fires threaten numerous structures and occupants in close proximity to wildland areas. A significant amount of time and resources are dedicated to reducing hazardous wildland fire fuel loads, which reduces the danger of a wildland fire.

Standards and codes: Expenditures on research and development have identified materials and designs that reduce the impact of a fire. These designs are frequently integrated into codes and standards, which require labor and resources to develop and maintain.

Insurance Premium: One method of mitigating the impact of fire is fire insurance, which requires annual (or semi-annual) policy premiums along with a paid deductible in the case of a fire. At the individual level this is a cost; however, at the society level much of this is a transfer where the premium is used to compensate for losses.

Fire prevention education/campaigns: A large percentage of fires are human caused; therefore, many fires can be reduced through behavior modification. Numerous organizations conduct education campaigns to inform the general public of the dangers of unintentional fires.

Aesthetic losses for fire protection design: Fire protection designs can require alteration of the appearance of a structure or product resulting in a loss of visual appeal.

Displacement costs (hotel and temporary shelter): Fire often renders residential structures as uninhabitable. The residence of these structures must find alternative shelter at their cost or their insurer's.

Fire code enforcement: For the sake of personal safety, many fire codes and standards are required by law; therefore, the government expends time and resources to enforce these requirements.

Fire incident investigation: A large number of fires are ignited intentionally, often for excitement, profit, or revenge; therefore, fire incidents are investigated to determine if a crime has been committed. Fires are also investigated to identify patterns of failure which warrant remediation or changes to codes and standards.

Summary of Losses

Direct property damage: Damage or destruction of public or private property directly resulting from a fire.

Indirect property damage: It is common for direct property damage to result in indirect loss such as loss due to business interruption or a decline in property value.

Economic value of civilian fatalities: The statistical value of life assigned to a civilian fatality.

Economic value of civilian injuries: The statistical value of life assigned to a civilian injury.

Economic value of fire fighter fatalities: The statistical value of life assigned to a fire fighter fatality.

Economic value of fire fighter injuries: The statistical value of life assigned to a fire fighter injury.

Environmental degradation: Fires burn residential and commercial facilities that contain hazardous materials. When these materials and others are consumed by fire they emit pollution into the air and can produce hazardous wastewater runoff.

Ecosystem and landscape degradation: Fires, especially wildfires, damage the natural environment as well as the stability of the landscape. Unstable landscapes can result in erosion, landslides, and other hazards.

Loss of recreational use: When the natural environment and recreational structures are damaged or destroyed they no longer can be used for the recreational uses they facilitated.

Timber loss: The timber value lost from forest fires.

Direct business interruption: Loss of income that a business suffers due to direct property damage.

Indirect business interruption: Loss of income due to direct property damage of nearby structures or infrastructure, including those of suppliers and customers.

Deadweight losses: The inefficiency resulting from the equilibrium of a good or service not being optimal.

Psychological losses (irreplaceable goods and peace of mind): The destruction of homes and businesses destroy personal goods with sentimental value. This destruction also has mental affects on individuals.

Transportation degradation: Smoke and fire often inhibits transportation or damages transportation routes.

Infrastructure losses: Damage to the physical structures and/or interruption of services necessary for the operation of society or enterprise.

Utility and infrastructure losses: The loss of water supply, sewers, electrical grids, and telecommunications services due to damage caused by fire.

Wildlife losses: Destruction of non-domesticated plants, animals, and other organisms due to fire.

Income losses: Lost income from missed work days due to non-injury circumstances.

Unfortunately, data is not available on all types of costs and losses. Furthermore, many of the current and more prominent values for costs and losses employ assumptions and techniques that result in imprecise estimates. These assumptions and techniques are necessary because of limitations of available data.

3.2 Stakeholders

Stakeholders are individuals, or groups of like individuals, who are affected by fires. They evaluate their costs and losses due to fire from their "stakeholder" perspective. Individual stakeholders may maintain different costs and losses than those maintained by a more aggregate perspective. For example, an individual stakeholder such as a homeowner has different costs and losses than the general public. It's important to understand the various burdens that fire incidents place on society and the economy, so to understand how changes in the fire problem (through a reduction in the associated costs and losses) are distributed. As seen in Table 3.2, there are public and private stakeholders with both costs and losses due to fire incidents. Homeowners and business owners are the major bearers of loss. Much of their losses, however, are transferred, in some sense, to the insurance industry through reimbursement of financial losses. However, homeowners and to a lesser extent businesses often have to bear the burden of sentimental and/or psychological losses.

Beyond the private ownership of homes and businesses, there are burdens that no single individual or subgroup bears on its own; these are burdens of the general public. The general public bears the burden of financing fire services and other emergency responders. It also bears the loss of life, the degradation of the landscape, and the loss of recreational use of public space. Transportation routes are often disturbed by smoke from wildfire incidents, electrical lines are damaged, wildlife is lost, and the environment is damaged. These losses affect the owners of damaged infrastructure, but also affect the general public that they serve.

Many local, state, and federal organizations are dedicated to the mitigation of fire incidents through the research and/or design of products and facilities. These include public as well as private interest groups such as the National Fire Protection Association or the National Institute of Standards and Technology. These organization's efforts come at some cost, however, as employees must be compensated for time and resources. These include the costs of standards and code development, product design and testing, and structural design and testing.

Table 3.2: Stakeholders of Fire Costs and Losses

Stakeholders	Affiliation	Costs	Losses
Homeowners	Private	Insurance Aesthetics Fire protection of facilities Fire protection goods and equipment Maintainance and repair of facility fire protection Displacement	Direct property loss Civilian deaths and injuries Psychological losses Income losses Loss of recreational use
Business Owners	Private	Insurance Aesthetics Fire protection of facilities Maintainance and repair of facility fire protection Fire protection goods and equipment	Direct property damage (net insurance) Indirect property damage (net insurance) Economic value of civilian fatalities Economic value of civilian injuries Direct business interruption Indirect business interruption
Government Interest Groups National Institute of Standards and Technology Consumer Product Safety Commission Occupational Safety and Health Administration National Interagency Fire Center U.S. Fire Administration Other	Public	Fire Protection Research and Design Fire prevention education/campaigns Standards and Codes	None

15

Table 3.2: Stakeholders of Fire Costs and Losses (continued)

Stakeholder	Type	Activities	Costs/Losses
Public Land Managers U.S. Forest Service Bureau of Indian Affairs Fish and Wildlife Service Bureau of Land Management Bureau of Reclamation National Park Service Municipal Land Management State Land Management Other	Public	Fire Protection Research and Design Fire prevention education/campaigns Fire service personell and training Fire service equipment Evacuation services and planning Fire protection of structures against WUI fires Fuel management Standards and Codes	Direct property damage Indirect property damage Infrastructure losses Fire fighter injuries and Fatalities
Private Interest Groups National Fire Protection Association National Association of State Foresters SILVIS Laboratories International Code Council Other	Private	Standards and Codes Fire Protection Research and Design Fire prevention education/campaigns	None
General Public	Public	Aesthetic losses for fire protection design Financing of Public Fire Services	Economic value of civilian fatalities Economic value of civilian injuries Environmental degradation of the atmosphere Degradation of the landscape Loss of recreational use Deadweight losses Psychological losses (irreplaceable goods) Transportation degradation Infrastructure losses Timber loss Utility losses Wildlife losses

Table 3.2: Stakeholders of Fire Costs and Losses (continued)

Stakeholder	Type	Costs	Losses
Municipal Fire Departments	Public	Fire service personell and training Fire service equipment Fire protection of structures against WUI fires Fuel management Fire prevention education/campaigns Fire code enforcement Fire incident investigation	Economic value of fire fighter fatalities Economic value of fire fighter injuries
State Fire Marshals	Public	Fire code enforcement Fire incident investigation	None
Police and EMS	Public	Evacuation services and planning Emergency responder personell and training Emergency responder equipment Fire code enforcement Fire incident investigation	None
Insurance Companies	Private	Insurance Administration Reimbursement for Losses	None
Product Manufacturers	Private	Fire Protection Research and Design Standards and Codes	None

17

4 Fire Cost and Loss Data

Data on fire related costs and losses are not fully developed into a single database. As seen in Table 4.1, one of the primary sources of data is the National Fire Incident Reporting System (NFIRS), which is a voluntary fire incident reporting program that provides numerous details on fire incidents (U.S. Fire Administration National Fire Data Center 2010). Since it is a voluntary system, it does not fully represent the fire problem. A second source of data is the NFPA Survey that provides national estimates of fire incidents within municipal jurisdictions; however, it does not provide as much incident-specific detail as the NFIRS system (Karter 2010). These two data sources can be combined to create detailed estimates that identify the causes and circumstances of fires as well as the extent of deaths, injuries, and property damage. Unfortunately, however, the combination of these two data sets does not produce a comprehensive look at all fire in the U.S. The NFPA Survey data is only for municipal fire departments; therefore, state and federal fire departments that respond to wildfires are excluded. Table 4.1 provides a list of data sources for costs and losses related to fire.

Table 4.1: Datasets Available

Dataset	Source	Years Covered	Costs	Losses	Coverage	Scale	Frequency
National Fire Incident Reporting System	U.S. Fire Administration	1980-1999 (version 4.x) 2001-2010 (version 5.x)	none	Direct Property Loss	Nationwide, for reporting fire departments	Fire Incident	Every Incident
USFS Fire Data	U.S. Forest Service	1950-2010	none	Acres Burned	USFS Land	Fire Incident	Every Incident
BIA Fire Data	Bureau of Indian Affairs	1972-2010	none	Acres Burned	BIA Land	Fire Incident	Every Incident
FWS Fire Data	Fish and Wildlife Service	1979-2010	none	Acres Burned	FWS Land	Fire Incident	Every Incident
BLM Fire Data	Bureau of Land Management	1972-2010	none	Acres Burned	BLM Land	Fire Incident	Every Incident
BOR Fire Data	Bureau of Reclamation	1972-2010	none	Acres Burned	BOR Land	Fire Incident	Every Incident
NPS Fire Data	National Park Service	1972-2010	none	Acres Burned	NPS Land	Fire Incident	Every Incident
Situation Reports	FAMWEB	1999-2010	none	Acres Burned	Wildland Fire Dispatch Center	Fire Incident	Every Incident
NFPA Survey of Fire Departments for U.S. Fire Experience	National Fire Protection Association	1977-2010	Number of Firefighters	Direct Property Loss	Survey of Fire Departments	National Aggregate	Annual
Census of Government Finances	U.S. Census Bureau	1992-2008	Local Fire Protection	none	Government Entities	National Aggregate	Annual
Construction Put in Place	U.S. Census Bureau	1993-2010	Construction Costs	none	Nationwide	National Aggregate	Annual
Construction Cost Data (online)	RSMeans	Current	Construction Costs	none	Nationwide	Component and Total Building	Annual
Insurance Data (online)	Insurance Information Institute	Current	Insurance	Direct and Indirect Losses	Nationwide	National Aggregate	Annual
Annual Survey of Manufactures	U.S. Census Bureau	1977-2009	Manufacturing Costs	none	Nationwide and Regional	Aggregate	Annual
Economic Census	U.S. Census Bureau	1967-2007	Manufacturing Costs	none	Nationwide and Regional	Aggregate	Quinquennial

4.1 Loss Data

Of the 15 executive departments in the federal government, four of them are involved with the collection and/or distribution of unprocessed fire data. In addition to these government departments, two private entities also collect and/or distribute unprocessed fire data. One of the most recognized collectors/distributers of data is the U.S. Fire Administration (USFA). As seen in Figure 4.1, it is an entity of the Department of Homeland Security's (DHS) Federal Emergency Management Agency (FEMA). The mission of the USFA is to "provide national leadership to foster a solid foundation for our fire and emergency services stakeholders in prevention, preparedness, and response." The USFA is authorized to gather, analyze, and standardize fire information and collects data through the National Fire Incident Reporting System (NFIRS). The National Fire Protection Association (NFPA), also shown in Figure 4.1, is another entity that is recognized for collecting and distributing fire data. It is a private nonprofit organization with the mission of reducing the "worldwide burden of fire and other hazards on the quality of life by advocating consensus codes and standards, research, training, and education." The NFPA gathers data through its annual survey of municipal fire departments. The USFA data and the NFPA data can be combined together to create a detailed data set for municipal jurisdictions.

As seen in Figure 4.1, a number of other entities are recognized for collecting wildland fire data; these include the U.S. Forest Service (USFS), an entity of the U.S. Department of Agriculture; Bureau of Indian Affairs (BIA), an entity of the Department of the Interior (DOI); Fish and Wildlife Service (FWS), an entity of DOI; Bureau of Land Management (BLM), an entity of DOI; Bureau of Reclamation (BOR), an entity of DOI; National Park Service (NPS), an entity of DOI; and the National Fire and Aviation Management Web Applications (FAMWEB). Each of these entities is shown in red in Figure 4.1 along with their parent organizations. FAMWEB is an interagency effort that includes a number of entities, including the USFS's Federal Aviation Management

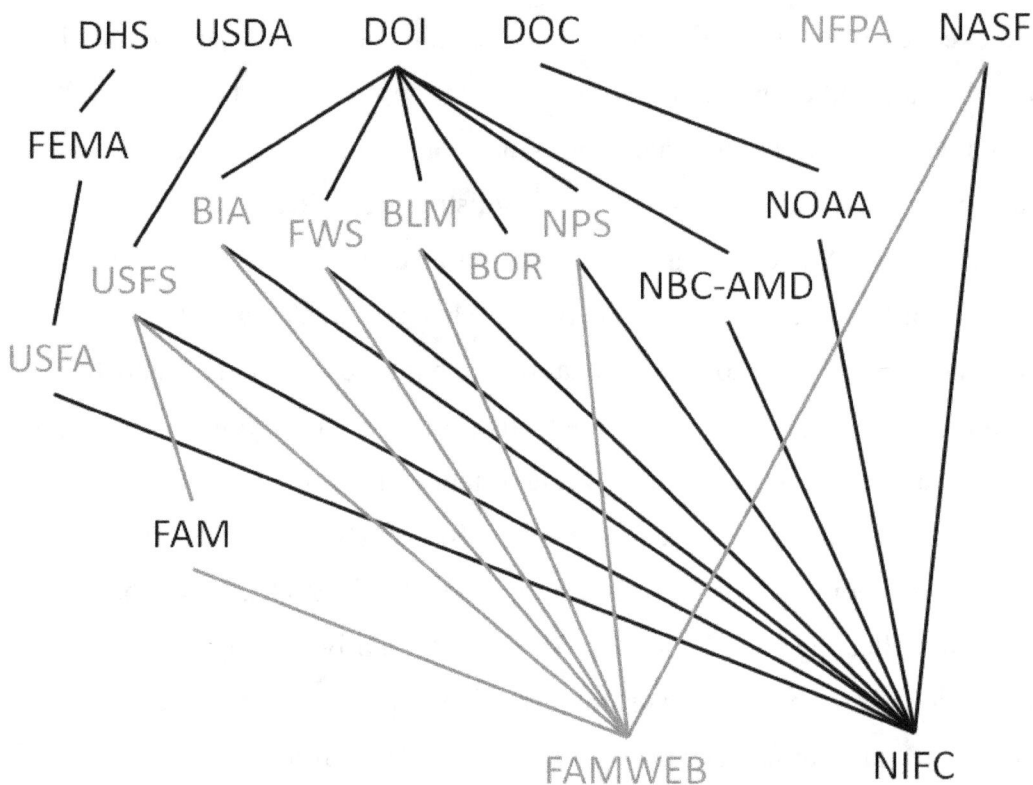

Figure 4.1: Organizational Chart of Entities Collecting and/or Distributing National Unprocessed Fire Data (distributers/collectors of unprocessed fire data are shown in red)

(FAM) and the National Association of State Foresters (NASF). The FAMWEB collects and distributes wildland fire data submitted by wildland fire dispatch centers through standardized forms. The National Interagency Fire Center (NIFC) is another interagency effort, and it includes the National Business Center's Aviation Management Directorate (NBC-AMD) and the Department of Commerce's (DOC) National Oceanic and Atmospheric Administration (NOAA).

The complexity of Figure 4.1 illustrates the need for a document discussing aggregate fire information. Additional complexity is added by the fact that data is collected in different formats. A number of the fire collection entities in Figure 4.1 have their own individualized method and/or format for collecting data. This dissimilarity makes the aggregation of fire data difficult. The following sections discuss the details of the unprocessed fire data so that aggregation might be possible.

4.1.1 National Fire Incident Reporting System

The National Fire Incident Reporting System (NFIRS) is a product of the National Fire Data Center, an entity of the U.S. Fire Administration (USFA). The USFA is authorized to gather, analyze, and standardize fire information and data through the Federal Fire Prevention and Control Act of 1974 (FEMA 2010). The reporting format in NFIRS is consistent with the National Fire Protection Association Standard 901, "Uniform Coding for Fire Protection." Approximately 1 000 000 total fires are reported to NFIRS each year by over 22 000 fire departments. USFA states that NFIRS is the world's largest, national, annual database of fire incident information; however, fire department participation in NFIRS is voluntary. Consequently, not all fire departments participate in NFIRS.

As described in the NFIRS reference guide, the NFIRS data system is broken into 11 modules[4] with the Basic Module as the primary one (FEMA 2010). In this module, general information is captured for every incident (emergency call). There are additional modules for different types of incidents. There is a Fire Module, Structure Fire Module, Civilian Fire Casualty Module, Fire Service Casualty Module, Wildland Fire Module, and an Arson Module. Information is generally obtained at the scene by emergency responder personnel. Within each module there are required fields and optional fields.

Not every module is relevant to all incidents (e.g., the *Structure Fire* module might not be relevant to an outside fire incident); therefore, some modules are mandatory (e.g., the *Basic Module*), while others are not (e.g., the *Arson Module*). Even within the mandatory modules, only some of the field elements are mandatory (e.g., Incident Type within the *Basic Module*). The *Basic Module* is completed for every emergency call to which the department responds. As seen in Table 4.2, it requests information about the location of

---

[4] The modules include the following: basic module, fire module, structure fire module, civilian fire casualty module, fire service casualty module, EMS module, hazardous materials module, wildland fire module, apparatus or resources module, personnel module, and the arson module.

the incident, incident type, alarm time and response time, actions taken (e.g., extinguishment), property losses, casualties, and property use. The column labeled 'primary identifiers' in Table 4.2, lists categories of the items that can be selected for the field category. Generally, there are a number of other items to select from under each of these categories. For example, the 'primary identifier' labeled structure fire in Table 4.2 has nine different types.

For non-contained[5] fires, the fire department is required to fill out the *Fire Module* or the *Wildland Fire Module*, which requests information about the ignition (e.g., area of origin, heat source, item first ignited), cause, and factors contributing to ignition, as seen in Table 4.3.[6] The *Structure Fire Module* is required for all structure fires. Structures can include "buildings, open platforms, bridges, roof assemblies, over open storage or process areas, tents, air-supported structures, and grandstands" (pg. 5-3, NFIRS 2006). As seen in Table 4.4, this module requests information about the structure type, building status, building height and floor size, fire origin, fire spread, and the presence and operation of smoke detectors and automatic fire sprinklers. There are additional fields in each of the modules above; however, most of them are optional fields. Entries recorded in NFIRS are listed based on incident and exposure numbers. One incident can have hundreds of exposures. For example, a wildland fire that spread throughout a neighborhood might have a unique exposure number for each house that caught fire, but all of them would have the same incident number. The total number of fires reported in the NFIRS data and the number of deaths and injuries is summarized in Table 4.5.

It is important for decision makers and community managers, who invest in mitigation techniques and strategies intended to minimize the cost and losses of fire, to know where fires are likely to occur and under what circumstances they are likely to ignite. Prioritizing mitigation investments to areas and circumstances with high fire probability and large values at-risk (also including losses of life and limb) requires a mechanism to

---

[5] Confined fires include fires confined to a cooking container, chimney or flue, incinerator, burner, compactor, or trash area.
[6] These modules are optional for confined fires, which include fires confined to a cooking container, chimney or flue, incinerator, burner, compactor, or trash area.

identify these areas. NFIRS provides detailed data that can be used to identify fires that might be prevented through engineering and behavioral solutions. Specifically, NFIRS provides the cause of a fire, the fuel and ignition source, and the location where these items were brought together. The NFIRS data, however, is vast and complex. The immense number of modules and fields makes analyzing NFIRS data difficult. In order to simplify this matter, the information can be categorized in a way that identifies areas where an engineering or behavioral solution might be appropriate.

For fire prevention purposes, four categories of causal detail have been developed as seen in Table 4.6. The categories begin with the most general cause of a fire and move toward specific details under which a fire occurs. The first category is the general causes of ignition; specifically, it is the NFIRS field labeled 'cause of ignition.' This field represents the general factor that resulted in a heat source igniting a combustible material. Both the category and the primary identifiers are displayed in the table. The next field is the specific causes of ignition. This includes the factors that allowed the condition or situation where the heat source and combustible material were combined to ignite a fire. The third category is the physical factors of ignition (e.g., heat source, item first ignited, and equipment involved in ignition). This category includes the combustible materials and the heat source involved in the ignition of a fire. The final category is the location characteristics of ignition (e.g., property use, area of origin, structure type, and structure status). This category includes the occupancy, location, type, and use of the property where the fire occurred and originated.

Table 4.2: Fields and Identifiers in the Basic NFIRS Module

Field	NFIRS Definition	Primary Identifiers
Incident Type	Situation Found	Structure fire; fire in mobile property used as a fixed structure; mobile property (vehicle) fire; natural vegetation fire; outside rubbish fire; special outside fire; cultivated vegetation, crop fire
Date, Time, and Address	Date and time of alarm and arrival along with the address of the incident	None
Actions Taken	Duties performed	Extinguish; salvage and overhaul; establish fire lines (wildfire); contain fire (wildland); confine fire (wildland); control fire (wildland)
Losses (property and contents)*	Total property and contents dollar loss	None
Deaths/Injuries	Deaths/injuries of civilians and fire fighters	None
Property Use	Use of the property where the incident occurred	Assembly; educational; health care, detention, and correction; residential; mercantile, business; industrial, utility, defense, agriculture, mining; manufacturing, processing; storage; outside or special property

* Not a required field

26

Table 4.3: Fields and Identifiers in the NFIRS Fire Incident Module

Module	Field	NFIRS Definition	Primary Identifiers
Fire Incident	Area of Origin	The primary use of the area where the fire started	Means of egress; assembly, sales areas; function area; technical processing area; storage area; service, equipment area; service, equipment area; structural area; transportation, vehicle area; other
Fire Incident	Heat Source	The heat source that ignited the item first ignited	Operating equipment; hot or smoldering object; explosives, fireworks; other open flame or smoking materials; chemical, natural heat sources; heat spread from another fire; other heat sources
Fire Incident	Item First Ignited	The item or material first ignited by the heat source	Structural component, finish; furniture, utensils, including built-in furniture; soft goods, wearing apparel; adornment, recreational material, signs; storage supplies; liquids, piping, filters; organic materials; general materials; general materials continued
Fire Incident	Cause of Ignition	The causal factor that resulted in ignition	Intentional; unintentional; failure of equipment or heat source; act of nature; cause under investigation; cause undetermined after investigation
Fire Incident	Factors Contributing to Ignition	Factors that allowed the heat source and material to ignite a fire	Misuse of material or product; mechanical failure, malfunction; electrical failure, malfunction; design, manufacturing, installation deficiency; operational deficiency; natural condition; fire spread or control
Fire Incident	Human Factors	Human condition or situation that allowed ignition	Asleep; possibly impaired by alcohol or drugs; unattended or unsupervised person; possibly mentally disabled; physically disabled; multiple persons involved; age was a factor

27

Table 4.4: Fields and Identifiers in the NFIRS Structure Fire Module

Module	Field	NFIRS Definition	Primary Identifiers
Structure Fire	Structure Type	Specific property type	Structure type, other; enclosed building; fixed portable or mobile structure; open structure; air supported structure; tent; open platform; underground structure work areas; connective structure
Structure Fire	Building Status	Operational status of the building involved	Under construction; occupied and operating; idle, not routinely used; under major renovation; vacant and secured; vacant and unsecured; being demolished; other
Structure Fire	Building Height	Number of stories above grade level	None
Structure Fire	Main Floor Size	Size of the main floor in square feet	None
Structure Fire	Fire Spread	Extent of flame damage	Confined to object of origin; confined to room of origin; confined to floor of origin; confined to building of origin; beyond building of origin
Structure Fire	Presence of Detectors	Existence of fire detection equipment within range of fire	None
Structure Fire	Presence of automatic extinguishing system (AES)	Existence of an AES within range of fire	None

Wildland Fires: The NFIRS data provides an opportunity to examine fires beyond those in structures and vehicles. A complete representation of the fire problem needs to include wildland fire activity. Currently, wildland fires within municipal jurisdictions are largely excluded from wildland fire statistics. The NFIRS data provides an opportunity to develop national estimates of the wildfires responded to by municipal fire departments; however, it is important to note that wildland fires within state and federal jurisdictions is largely absent from NFIRS data. NFIRS is one of the few wildland fire data sets that provides dollar estimates for losses. This information can be used by wildland managers and researchers to analyze the economic returns to wildfire suppression, fuels management, and prevention activities. Furthermore, the presence of wildland fires within municipal jurisdictions implicate that there is a need for forest management beyond state and federal lands. Specifically, there is a need for forest management within municipal jurisdictions.

Table 4.5: Fires and Losses Reported in NFIRS

Year	Fires	Property Loss ($billion)	Contents Loss ($billion)	Fire Fighter Deathes	Civilian Deaths	Fire Fighter Injuries	Civilian Injuries
2001	620 807	3.3	0.5	6	1 478	5 759	7 939
2002	700 564	3.2	0.7	24	1 504	5 702	8 141
2003	873 089	3.5	0.9	17	1 890	6 162	9 486
2004	934 640	3.7	1.2	27	1 906	6 485	9 474
2005	1 126 792	4.3	1.4	34	1 913	7 259	9 656
2006	1 211 630	5.7	1.7	30	2 075	7 415	9 771
2007	1 219 301	6.7	2.2	39	2 020	8 360	9 854
2008	1 157 846	6.9	2.4	26	2 086	9 907	9 840
2009	1 111 888	5.5	2.0	17	1 930	9 956	9 670
Total	8 956 557	43	13	220	16 802	67 005	83 831

Source: U.S. Fire Administration National Fire Data Center. (2010) National Fire Incident Reporting System 5.0. FEMA, Washington, DC.

Table 4.6: NFIRS Categories by Causal Detail for Fire Prevention Purposes

Location Characteristics of Ignition

Property Use
- Assembly
- Educational
- Health Care, Detention & Correction
- Residential
- Mercantile, Business
- Industrial, Utility, Defense, Agriculture, Mining
- Manufacturing, processing
- Storage
- Outside or special property
- Other

Area of Origin
- Means of Egress
- Assembly, Sales Areas (Groups of People)
- Function Area
- Technical Processing Areas
- Storage Areas
- Service Areas
- Service, Equipment Areas
- Structural Areas
- Transportation, Vehicle Areas
- Other Area of Origin

Structure Type
- Structure type, other
- Enclosed building
- Fixed portable or mobile structure
- Open structure
- Air supported structure
- Tent
- Open platform
- Underground structure work areas
- Connective structure

Structure Status
- Under construction
- Occupied and operating
- Idle, not routinely used
- Under major renovation
- Vacant and secured
- Vacant and unsecured
- Being demolished
- Other

Physical Factors of Ignition

Equipment Involved in Ignition
- Heating, Ventilating & Air Conditioning
- Electrical Distribution, Lighting & Power Transfer
- Shop Tools & Industrial Equipment
- Commercial & Medical Equipment
- Garden Tools & Agricultural Equipment
- Kitchen & Cooking Equipment
- Electronic and Other Electrical Equipment
- Personal & Household Equipment
- Other

Item First Ignited
- Structural Component, Finish
- Furniture, Utensils, including built-in furniture
- Soft Goods, Wearing Apparel
- Adornment, Recreational Material, Signs
- Storage Supplies
- Liquids, Piping, Filters
- Organic Materials
- General Materials
- General Materials Continued
- Other

Heat Source
- Operating equipment
- Hot or Smoldering Object
- Explosives, Fireworks
- Other Open Flame or Smoking Materials
- Chemical, Natural Heat Sources
- Heat Spread from Another Fire
- Other Heat Sources

Specific Causes of Ignition

Human Factors Contributing to Ignition
- Asleep
- Possibly impaired by alcohol or drugs
- Unattended or unsupervised person
- Possibly mentally disabled
- Physically disabled
- Multiple persons involved
- Age was a factor
- Other

Factors Contributing to Ignition
- Misuse of Material or Product
- Mechanical Failure, Malfunction
- Electrical Failure, Malfunction
- Design, Manufacturing, Installation Deficiency
- Operational Deficiency
- Natural Condition
- Fire Spread or Control

General Causes of Ignition

Cause of Ignition
- Intentional
- Unintentional
- Failure of equipment or heat source
- Act of nature
- Cause under investigation
- Cause undetermined after investigation
- Other

4.1.2 Federal Land Manager Data

A number of federal entities have their own databases for wildfires. The Bureau of Indian Affairs, Bureau of Land Management, Forest Service, Fish and Wildlife Service, and the National Park Service each have their own databases for tracking wildfires (National Fire and Aviation Management 2011). Fortunately, a software program called Fire Family Plus (Missoula Fire Sciences Laboratory 2008) has been developed to bring these data sets together. The data is available through the National Fire and Aviation Management website (FAMWEB).

There are two formats for the data imported through Fire Family Plus. The first format is for data available from DOI entities. This format includes various fire types as listed in Table 4.7; some of these fire types overlap. For example, a data set that included 'agency land / other federal suppression' and 'other land / agency suppression – agreement' for all DOI entities could count some fires twice. This double counting occurs because one agency could report a fire as 'agency land / other federal suppression' while another agency reported the same fire as 'other land / agency suppression – agreement.' This same situation occurs with the 'agency land / natural out / other federal protection' and 'other land / natural out / agreement' fire types. It is for this reason that 'agency land / other federal suppression,' 'agency land / natural out / other federal protection,' and 'other land / agency assist' have been excluded from the statistics in Table 4.8 that summarize the federal land manager data. Prescribed burns are also excluded, since these do not represent part of the fire problem, but part of the costs.

Table 4.7: Federal Fire Types (DOI entities)

Department of Interior Fire Types	Overlap
Agency Land / Agency Suppression	
Agency Land / Other Federal Suppression	*
Agency Land / Non-Federal Suppression	
Agency Land / Confine or Contain	
Other Land / Agency Suppression / Threat to Agency	
Other Land / Agency Suppression - Agreement	*
Agency Land / Appropriate Management Response	
Agency Land / Natural Out / Agency Protection	
Agency Land / Natural Out / Other Federal Protection	**
Agency Land / Natural Out / Non-Federal Protection	
Other Land / Natural Out / Threat to Agency	
Other Land / Natural Out / Agreement	**
Other Land / Agency Assist	
Prescribed Burn / Within Prescription	
Prescrived Natural / Within Prescription	

* Overlapping fire types

** Overlapping fire types

Table 4.8: Aggregate Fire Data Reported by Federal Land Managing Entities

	2002	2003	2004	2005	2006	2007	2008	2009	2010
Natural	9 654	12 141	9 873	8 294	12 730	8 983	6 562	7 716	5 820
Campfire	1 347	1 389	1 350	1 535	1 845	1 510	1 062	915	972
Smoking	615	382	290	329	354	309	194	166	163
Incendiary	1 686	1 589	1 746	1 411	1 346	1 217	1 207	964	1 061
Equipment	103	87	91	103	117	97	80	47	61
Railroads	2 466	2 308	2 200	3 507	4 221	3 574	2 968	2 483	2 344
Juveniles	1 060	1 265	777	990	1 449	1 026	753	602	572
Miscellaneous	3 871	3 065	2 998	3 706	4 071	3 919	2 424	2 435	2 608
TOTAL	20 802	22 226	19 325	19 875	26 133	20 635	15 250	15 328	13 601

Source: Fire Family Plus v. 4.0. September 2008. <http://www.firemodels.org/index.php/national-systems/firefamilyplus>

4.1.3 Incident Management Situation Report

Incident Management Situation Reports (IMSR) are developed from Incident Command System (ICS) reports; specifically, they are the Incident Status Summary or ICS-209 forms filled out primarily by wildland fire dispatch centers. ICS is a standardized all-hazards management approach. It is used by all levels of government, including Federal, State, tribal, and local governments. ICS-209 information is intended to be submitted by the closest dispatch center (National Interagency Fire Center 2011) to the incident and the data is maintained on an Oracle database. IMSR data is provided to the public through the National Fire and Aviation Management website (FAMWEB). At the beginning of each year the Oracle database is published as a Microsoft Access file available for download on the website. The data is also accessible by geographic region directly on the FAMWEB website. Unfortunately, there is no manual or guide that explains the Microsoft Access data, variables, or relationships between the various tables in the IMSR data; however, there is a manual for the data available directly from the FAMWEB website.

Although the Microsoft Access files are based on the Oracle database, there is a discrepancy between the two data sets. The difference between the data sets can be significant and the cause for the discrepancy is not documented or readily apparent to data users. In the past, the data was entered and stored as an Oracle database. In January of each year, NIFC staff generated estimates from the Oracle database, which were later published on the NIFC website. Later in the year, Microsoft Access files were generated from the Oracle database; these are the files that are available on the FAMWEB website. Between the time that the NIFC estimates were generated and the time that the Access files were generated, fire management personnel had the ability to change the figures in the Oracle database. This editing of the data created a discrepancy between the source data and the NIFC estimates. As of January 2011, fire management personnel cannot change the figures in the Oracle database beyond January 1; thus, 2010 NIFC figures and 2010 figures from the FAMWEB Microsoft Access files match while data from previous years does not match.[7] Since the difference between these datasets is not insignificant, it is important to be aware of this discrepancy when tracking the wildfire problem.

Table 4.9 provides a list of the various agencies that submit wildland fire data to IMSR. The majority of the agencies listed are federal agencies; however, state agencies represent a large proportion of the fires reported. The IMSR fire statistics data contains the number of fires and acres burned by fire unit by cause, including human caused and lightning

---

[7] This information was gathered through correspondence with NIFC staff.

Table 4.9: Agencies that Submit Data to the IMSR

Agency Abbreviation	IMSR Agency	Authority
APHI	Animal/Plant Health Insp. Serv	Federal
BIA	Bureau of Indian Affairs	Federal
BLM	Bureau of Land Management	Federal
CNTY	County	County
DDQ	Department of Defense	Federal
DHS	Depart. of Homeland Security	Federal
FWS	Fish and Wildlife Service	Federal
IA	Interagency	Federal
INTL	International	Other
LGR	Local Government Resource	Local
NPS	National Park Service	Federal
OES	Office of Emergency Services	State
ORC	Orange Co. Fire Authority	County
OTHR	Other	Other
PRI	Private Contractors	Private
ST	State Government Agencies	State
USFS	U.S. Forest Service	Federal
WXW	National Weather Service	Federal

Table 4.10: Wildland Fires Reported to the IMSR

Year	Human Caused		Lightning Caused		TOTAL	
	Fire Count	Acres	Fire Count	Acres	Fire Count	Acres
2002	52 787	2 696 687	10 613	3 498 554	63 400	6 195 241
2003	50 847	1 924 026	12 761	2 037 948	63 608	3 961 974
2004	54 167	955 173	11 372	7 006 833	65 539	7 962 006
2005	58 372	1 513 546	8 330	7 168 136	66 702	8 681 682
2006	56 903	3 542 586	11 420	4 814 435	68 323	8 357 021
2007	72 297	3 401 272	11 835	5 844 533	84 132	9 245 805
2008	55 336	2 963 013	7 861	1 668 031	63 197	4 631 044
2009	53 055	1 716 959	7 958	3 813 712	61 013	5 530 671
2010	64 810	1 303 490	7 164	2 119 239	71 974	3 422 729
Average	57 619	2 224 084	9 924	4 219 047	67 543	6 443 130

Source: National Interagency Fire Center. 2011 National Fire and Aviation Management Web Applications (FAMWEB). SIT Reports. 2002-2010. <https://fam.nwcg.gov/fam-web/>

caused fires. Table 4.10 contains the count of fires reported by fire units from each unit's final year to date (YTD) report in the IMSR for 2002 through 2010. The average number of wildfires for this time period is approximately 68 thousand.

4.1.4 National Fire Protection Association

NFPA data is from an annual survey where 3000 fire departments are randomly selected from 30 300 municipal fire departments to produce nationwide estimates of residential and non-residential structure fires, vehicle fires, and outside and other fires (Karter 2010). The NFPA places fire departments in one of 10 strata by size of community protected:

1 000 000 and up

500 000 to 999 999

250 000 to 499 999

100 000 to 249 999

50 000 to 99 999

25 000 to 49 999

10 000 to 24 999

5 000 to 9 999

Under 2 500

The sample sizes for each strata are selected to provide the best estimate of losses for one-and two-family dwellings. All departments that protected 100 000 or more people were included. In addition to numbers of fires, the NFPA survey collects data on damages and civilian (non fire service) injuries and fatalities (Karter 2010). While the survey provides a rigorous method for estimating high-level statistics related to fires that local municipal fire departments responded to, it does not provide detailed insight into the specifics of the incidents. Table 4.11 provides the estimated number of fires calculated using the NFPA survey.

Table 4.11: NFPA National Fire Estimates for Municipal Fire Departments

	Fires	Civilian Deaths	Civilian Injuries	Firefighter Deaths	Firefighter Injuries	Direct Property Damage (billions current dollars)	Direct Property Damage (Billions 2009)
2002	1 687 500	3 380	18 425	97	80 800	10.3	12.3
2003	1 584 500	3 925	18 125	106	78 750	12.3	14.4
2004	1 550 500	3 900	17 875	105	75 840	9.8	11.1
2005	1 602 000	3 675	17 925	87	80 100	10.7	11.8
2006	1 642 500	3 245	16 400	89	83 400	11.3	12.0
2007	1 557 500	3 430	17 675	103	80 100	14.6	15.1
2008	1 451 500	3 320	16 705	104	79 700	15.5	15.5
2009	1 348 500	3 010	17 050	82	78 150	12.5	12.5
Average	1 553 063	3 486	17 523	97	79 605	12.1	12.7

Source: National Fire Protection Association. 2011. "The U.S. Fire Problem." <http://www.nfpa.org>

4.1.5 National Interagency Fire Center

The National Interagency Fire Center generates aggregate wildland fire data that is used by the USFA (National Interagency Fire Center 2011). These estimates are generated solely using the last year-to-date report submitted by fire units listed in the IMSR data, which is available on the FAMWEB website. As mentioned previously, there was a discrepancy between NIFC data and the source data from the FAMWEB. This information is important since the data discrepancy is not documented and the difference in values is not insignificant. As of January 2011, the data management process has changed; thus, 2010 NIFC figures and 2010 figures from the FAMWEB Microsoft Access files are nearly identical. Data from previous years does not match. The number of wildland fires estimated by NIFC between 2003 and 2010 is shown in Table 4.12 and the number of acres burned is shown in Table 4.13.

Table 4.12: Fires by Agency Reported by the NIFC

	2003 Fires	2004 Fires	2005 Fires	2006 Fires	2007 Fires	2008 Fires	2009 Fires	2010 Fires
BIA	4 095	3 662	5 110	6 768	4 593	4 934	4 375	3 825
BLM	2 931	2 906	2 655	3 848	2 613	1 941	2 545	2 312
CNTY	1 011	483	609	1 201	835	15 429	15 467	6 348
DDQ	40	71	39	83	333	150	126	348
FWS	358	382	518	524	398	425	448	323
NPS	486	490	394	537	489	396	426	390
OTHR	148	119	345	337	390	266	349	297
PRI	1	1	764	2 017	12 641	5 512	1 185	8 321
ST	44 263	48 739	48 844	70 667	54 927	42 680	46 179	43 010
USFS	10 258	8 608	7 268	10 403	8 486	6 930	7 691	6 797
TOTAL	63 591	65 461	66 546	96 385	85 705	78 663	78 791	71 971

Source: National Interagency Fire Center. 2011. "Historical Wildland Fire Summaries." 2003-2010.
<http://www.nifc.gov/fireInfo/fireInfo_statistics.html>

Table 4.13: Acres burned by Agency Reported by the NIFC

	2003 Acres	2004 Acres	2005 Acres	2006 Acres	2007 Acres	2008 Acres	2009 Acres	2010 Acres
BIA	269 777	71 292	194 733	376 824	266 593	168 336	200 562	106 978
BLM	352 466	1 305 794	3 591 721	2 406 622	2 021 009	330 981	989 029	830 377
CNTY	81 928	29 653	107 427	345 830	178 235	723 277	443 413	284 626
DDQ	6 336	24 356	73 649	13 021	25 454	66 545	69 971	63 037
FWS	325 515	2 096 403	1 842 229	236 746	501 044	95 952	821 838	187 991
NPS	196 895	42 352	128 824	73 566	102 459	89 061	182 047	174 255
OTHR	69 889	911 065	260 326	285 040	308 907	16 777	9 175	227 991
PRI	0	1	71 256	472 242	425 722	176 759	19 641	136 062
ST	1 229 639	3 064 998	1 636 995	3 767 793	2 663 045	2 351 938	2 470 433	1 091 677
USFS	1 428 283	551 966	779 593	1 896 071	2 835 577	1 223 817	715 677	319 730
TOTAL	3 960 728	8 097 880	8 686 753	9 873 755	9 328 045	5 243 443	5 921 786	3 422 724

Source: National Interagency Fire Center. 2011. "Historical Wildland Fire Summaries." 2003-2010.
<http://www.nifc.gov/fireInfo/fireInfo_statistics.html>

4.1.6 Detailed Fire Count Methodology

The NFPA uses a methodology developed by Hall and Harwood (1989) to provide national estimates of fire. The method multiplies the number of reported fires from NFIRS by a scaling ratio for four different types of fires: residential, nonresidential, vehicle, and outside and other fires. The procedure can be defined by the following equation:

$$E_{i,j} = R_{i,j} * \frac{S_{i,j}}{R_{i,j}}$$

where E is the national estimate of j (j is either the number of fires, losses, fatalities, or injuries) for incident type i (i is either residential structure fires, non-residential structure fire, vehicle fire, or outdoor and other fire). Variable R is the total reported value (from NFIRS) of j for incident type i. Variable S is the NFPA (surveyed) estimate of j for incident type i. Variable R is total reported value (from NFIRS) of j of incident type i. For example, to calculate the national estimate (E) for the number of residential structures (j), take the number of residential structure fires in NFIRS and multiply it by the number of residential structure fires estimated by the NFPA divided by the number of residential structure fires recorded in NFIRS. This procedure is used extensively in documents published by USFA and the NFPA.

Since the NFIRS data is scaled using NFPA survey data obtained from municipalities, any data in NFIRS reported by the National Park Service, U.S. Forest Service, Bureau of Indian Affairs, Bureau of Land Management, U.S. Fish and Wildlife Service, state department data, and other non-municipal data is deleted from the NFIRS data (this comprised only a few incidents). Data with missing values is also deleted. Deleting data with missing values assumes that the unlabeled variables are distributed similar to the labeled variables.

4.1.7 Data Integration

Some of the fire data sets discussed in the previous sections provides fire data for multiple jurisdictions; however, each data set may represent a particular jurisdiction better than the other data sets. The NFIRS data primarily represents municipal fire departments. It does contain some non-municipal data (e.g., state and federal fire departments); however, this is largely under represented. Additionally, municipal data from NFIRS can be joined with municipal data from the NFPA to provide accurate municipal fire estimates. These two data sets are the only ones that provide dollar estimates for losses. Federal land manager data solely represents federal wildland fire jurisdictions and the IMSR data focuses on wildland fire dispatch centers, which can include any wildland jurisdiction. These two data sets provide the number of fires and the number of acres burned, but do not provide any dollar estimates for losses. The IMSR data tends to under represent wildland fires in municipal jurisdictions since these jurisdictions generally do not have wildland fire dispatch centers. This is a significant factor since municipal jurisdictions encompass the majority of the U.S. population. That is, both the primary cause of fires (humans) and a large percent of the value-at-risk are contained in these jurisdictions. In terms of federal wildland jurisdictions, the data that is provided by the federal land managers provides greater detail on the source of the data than does the IMSR data.

For the nine entities that provide unprocessed fire data, there are five formats used for data collection/distribution: NFIRS data format, USFS data format, DOI data format, IMSR data format, and the NFPA survey format. Therefore, it is essential to allocate attention to the type and format of data included in each data set if they are to be combined. One data set may overlie another data set.

4.2 Cost Data

With the limitations of fire loss data, the data associated with fire costs is, in many ways, more limited. Many expenditures and costs are bundled with other costs; for example, insurance policies are bundled into multi-hazard policies and fire research expenditures are bundled with all types of research. There are few instances where data is specifically collected for costs related to fire incidents. The following sections discuss the limited data that is available on costs related to fire incidents. In some cases techniques are used to estimate costs or losses; since this report focuses on data availability, these techniques are discussed only briefly.

4.2.1 Insurance Data

Many losses resulting from fire are insured; however, the insurance industry data is limited and scarcely available to the public. Insurance cost estimates must take into account both the premium paid to insurers and claims made by victims of fire incidences. Insurance data is complicated due to the fact that each insurance company contains their own data and this data is not comprehensively published; for example, claims paid to victims are not published. Additionally, most insurance policies are not single hazard policies; that is, an insurance policy generally covers multiple hazards (i.e., multiple peril policy); therefore, singling out the cost of one hazard (fire) requires some investigation as well as some assumptions.

The primary source of insurance data is the Insurance Information Institute, which is a U.S. industry organization developed to improve public understanding of insurance. Its primary function is to provide accurate and timely information on insurance subjects. The Insurance Information Institute's Insurance Fact Book (2011) provides data that can be used to make estimates of fire insurance expenditures. This annually published document provides facts and figures on property/casualty and life/health insurance. It provides expenditures on insurance as well as insured losses. The estimates for total direct premiums written by line for property/casualty insurance in 2009 include the following: farm owners multiple peril ($2.82 billion), commercial multiple peril ($33.33 billion), homeowners multiple peril ($67.44 billion), and commercial fire ($12.08 billion). In 2008, fire and lightning represented 27.38 % of all homeowner claims (Insurance Information Institute 2011). Assuming that 27.38 % of all the multiple peril policies (farm owner, commercial, and homeowner multiple peril) are attributed to fire, it can be estimated that approximately $28.4 billion of the total $103.6 billion paid in multiple peril premiums is attributed to fire. The total premium attributed to fire is then estimated to be $40.4 billion ($28.4 billion in multiple peril plus $12.1 billion in commercial fire). The Insurance Fact Book also estimates that approximately 26 % of revenue was spent on expenses while 74 % was spent on claims (Insurance Information Institute 2011). Assuming that 26 % of all fire premiums are attributed to expenses, it can be estimated that expenses attributed to fire insurance (premiums minus claims for losses) was $10.5 billion in 2009.

Beyond the Insurance Information Institute, specific fire insurance data is limited. Insurance companies are hesitant to publish proprietary data since it may provide their competition with an advantage. Additionally, there are few incentives to expend resources doing so.

4.2.2 Emergency Responder Personnel and Equipment Data

Expenditures: The U.S. Census Bureau estimates state and local fire protection estimates annually using its Census of Government Finances. The most recent data released is for 2007; the total expenditure on state and local fire protection for this year is estimated to be $36.828 billion (U.S. Census Bureau 2011), as seen in Table 4.14. This includes "fire fighting organization and auxiliary services; fire inspection and investigation; support of volunteer fire forces; and other fire prevention activities [and also] includes [the] cost of fire fighting facilities, such as fire hydrants and water, furnished by other agencies of the government" (U.S. Census 2007). It is assumed by Hall (2011) that this includes expenditures on career firefighters, purchases of fire protection equipment, and other related items. Excluded is the monetary value of volunteer firefighters. It is also important to note that although fire departments primary function is to respond to fires they also respond to emergencies other than fires; for example, they frequently respond to medical emergencies. The Census Bureau also provides estimates for police protection; however, it does not provide data on expenditures related specifically to fire incidents and investigations.

Table 4.14: Local Government Expenditures on Fire Protection (current dollars, thousands)

Year	Local Government
2007	36 827 892
2006	33 654 566
2005	30 830 112
2004	28 990 555
2003	27 854 042
2002	25 997 621
2001	24 970 060
2000	23 101 931

Source: U.S. Census Bureau. 2011. "The 2011 Statistical Abstract: State & Local Government Finances & Employment: Revenue and Expenditures by Function." <http://www.census.gov/compendia/statab/>

Total state and federal expenditures on fire response are more difficult to assess. The Census Bureau does not provide estimates for either of these categories. Although costs are often assessed for individual fire incidents, aggregate expenditure data is not readily available. Hall (2011) suggests that these costs are typically small when

41

compared to the expenditures by local governments.

Volunteer Firefighters: Because there are considerably more volunteer firefighters than career, the value of their donated time is a significant cost to society. Specific data on the time these individuals donate is not available on an annual basis; however, data on the number of volunteer fire fighters is available through the NFPA Survey for U.S. Fire Experience and the NFPA Fire Service Survey, as displayed in Table 4.15 (Karter and Stein 2010). Using this data, there were 2.42 volunteer firefighters for every career firefighter in 2009. The number of hours that each volunteer donates is not fully documented, nor the value they place on their volunteered time; however, some organizations, such as the Independent Sector (2011) do provide estimates of the value of volunteer time (by state and for the U.S.).

Table 4.15: Number of Firefighters in the U.S., 2000-2009

Year	Volunteer Firefighters	Career Firefighters	Total
2000	777 350	286 800	1 064 150
2001	784 700	293 600	1 078 300
2002	816 600	291 650	1 108 250
2003	800 050	296 850	1 096 900
2004	795 600	305 150	1 100 750
2005	823 350	313 300	1 136 650
2006	823 950	316 950	1 140 900
2007	825 450	323 350	1 148 800
2008	827 150	321 700	1 148 850
2009	812 150	335 950	1 148 100

Source: Karter, Michael J. and Gary P. Stien. 2010. U.S. Fire Department Profile through 2009. NFPA. Quincy, MA.

4.2.3 Fire Protection Planning and Research and Development Data

Disaster Recovery Planning: Meade (1991a) identifies that there are costs associated with the development and implementation of disaster recovery plans. Data on recovery planning is limited; however, he estimates this cost at $680 million ($425 million for suppression systems and $255 million for disaster recovery planning). The estimate for suppression assumes that 2.2 % of the market for mainframes, super computers, and mini-super-computers ($19.4 billion) was for suppression. Disaster recovery was estimated via dialog with industry experts. Sources are limited beyond this estimate.

Research and Development: A number of public and private organizations participate in fire protection research and development. The Consumer Product Safety Commission, for example, provides a number of reports on fire incidents and maintains a database of product incidents. The National Institute of Standards and Technology also expends resources researching fire. There is not, however, a comprehensive database on these types of expenditures. The National Science Foundation and the Economic Census both track research and development expenditures, but there is not a breakout for fire protection expenditures. Meade (1991a) provides an estimate for fire retardant and flammability testing ($2.5 billion), which is based on dialog with industry experts. There is also a 1977 publication of the National Academy of Sciences' National Materials Advisory Board entitled 'Materials and Process Specifications and Standards' that estimates that $320 million was spent on preparing and maintaining standards. Meade (1991a) estimates that 25 % of this value as being related to fire. Beyond this estimate, there are few sources for data on expenditures for standards development. The budgets of some entities are available to the public; however, assembling expenditure data for all entities would require a significant amount of time and resources.

4.2.4 Data on the Implementation of Fire Protection Measures

Fire Protection in Equipment and Goods: Numerous products have been designed or redesigned with the risk of fire taken into account. Some cigarettes, for example, are designed as Fire Safe Cigarettes (FSC) that extinguish more quickly than standard cigarettes and automobiles often have heat shields over catalytic converters and other heated components to prevent fires. These types of design factors often require additional resources to produce and install. Data on these costs are largely missing. Each product has the potential to pose a unique fire risk and, therefore, requires a unique solution with associated costs. Collecting data on these costs requires a substantial amount of resources; as a result, specific data is unavailable. Meade (1991a), however, provides a onetime estimate for meeting "fire grade" standards in the manufacture of equipment ($18 billion) with the total product being valued at $78 billion. This is calculated as 40 % of the purchases of information processing and related equipment plus 40 % of the purchases for industrial equipment. Approximately 30 % or $18 billion is considered to be the premium paid for fire grade products. This assessment is based on dialog with industry experts along with an input-output analysis appearing in the Survey of Business. The percentages used in Meade (1991a) could be applied to modern estimates of manufacturing sales found in the Annual Survey of Manufactures.

Installation of Fire Protection in Constructed Facilities: Data on the expenditures for fire protection measures in constructed facilities is not collected in any substantial way; however, previous research has provided a method to estimate the cost of fire protection in constructed facilities. Research produced by three students at Worcester Polytechnic Institute (WPI) published in 1978 determined that 2.5 % of residential construction, 9.0 % of private nonresidential construction, 3.0 % of other private construction, and 4.0 % of public construction is attributed to fire protection in constructed facilities. The WPI research used estimates from the National Construction Estimator and Means Building System Cost Guide as insight into estimating the construction costs for fire related materials and products. Although building construction has changed somewhat from when these estimates were made 30 years ago, there are simply few alternative estimates and it is believed that they still represent an acceptable approximation. The WPI estimates continues to be cited in a number of publications including NFPA's 'Total Cost of Fire in the United States' (Hall 2011) and NIST's 'First Pass at Computing the Cost of Fire Safety in a Modern Society' (Meade 1991a), which was also published in the

Fire Technology Journal (Meade 1991b). Meade, however, references a higher percentage (12 %) for private nonresidential construction, which is used by both Meade and Hall. Fire protection costs estimated using the WPI method are shown in Table 4.17 along with the values for construction put in place. General cost data on construction is available from R.S. Means and other organizations; however, using this data to estimate the cost of fire protection in constructed facilities would a significant amount of expertise in the construction of various building types. As indicated by the WPI report, this type of estimation is complex (Apostolou et al, 1978).

Meade (1991a) points out that there are costs associated with the maintenance of fire protection and suppression equipment installed in factories and office buildings. This cost includes system maintenance, industrial fire brigades, and training programs for occupational fire protection and safety. Citing dialog with industry experts, Meade estimates that between 0.5 % and 2 % of manufacturing costs are attributed to this maintenance and that 50 % of sales is due to manufacturing; however, it is unnecessary to estimate manufacturing's percent of sales. The 2009 Annual Survey of Manufactures provides the total value added for manufacturing, which is the measure of manufacturing activity less the cost of materials, supplies, containers, fuel, purchased electricity, and contract work. As seen in Table 4.16, this works out to be 44.6 % of net sales (measured as total value of shipments) in 2009. It is important to note that the higher limit estimated by Meade (2 %) for maintenance of fire protection and suppression equipment installed in factories and office buildings exceeds the total value of maintenance and repair for manufacturing structures and machinery cited in the Annual Survey of Manufactures ($36 billion). Meade estimates that non-manufacturing costs for maintenance of fire protection and suppression equipment is one-fourth that of manufacturing costs.

Table 4.16: Manufacturing Shipments, Value Added, and Fire Prevention Expenditures (thousands of current dollars)

	Total Value of Shipments	Value Added	Value Added as a Percent of Shipments	Fire Prevention (0.5 % of Value Added)	Fire Prevention (2 % of Value Added)
2009	4 436 196 105	1 978 017 343	44.6 %	9 890 087	39 560 347
2008	5 468 093 135	2 266 362 722	41.4 %	11 331 814	45 327 254
2007*	5 319 456 312	2 382 643 001	44.8 %	11 913 215	47 652 860
2006	5 015 553 256	2 285 928 967	45.6 %	11 429 645	45 718 579
2005	4 742 076 879	2 210 349 247	46.6 %	11 051 746	44 206 985
2004	4 265 784 041	2 031 438 811	47.6 %	10 157 194	40 628 776
2003	4 015 080 847	1 926 388 030	48.0 %	9 631 940	38 527 761
2002	3 914 623 710	1 888 050 250	48.2 %	9 440 251	37 761 005
2001	3 967 698 457	1 850 709 351	46.6 %	9 253 547	37 014 187
2000	4 208 582 047	1 973 622 421	46.9 %	9 868 112	39 472 448

* 2007 data was taken from the Economic Census
Source: U.S. Census Bureau. 2011 "Annual Survey of Manufactures."
<http://www.census.gov/manufacturing/asm/index.html>

Table 4.17: Construction Put in Place and Fire Protection Costs (current dollars, millions)

	Percent Fire Protection	2006		2007		2008		2009		2010	
		Put in Place	Fire Protection	Put in Place	Fire Protection	Put in Place	Fire Protection	Put in Place	Fire Protection	Put in Place	Fire Protection
Private Residential Structures	2.5%	613 731	15 343	493 246	12 331	350 257	8 756	245 912	6 148	238 801	5 970
Private Nonresidential Structures	**9.0%**	222 531	20 028	267 170	24 045	290 532	26 148	227 763	20 499	159 088	14 318
Lodging	9.0%	17 624	1 586	27 481	2 473	35 364	3 183	25 388	2 285	10 904	981
Office	9.0%	45 680	4 111	53 815	4 843	55 502	4 995	37 282	3 355	24 231	2 181
Commercial	9.0%	73 368	6 603	85 858	7 727	82 654	7 439	50 460	4 541	37 647	3 388
Health care	9.0%	32 016	2 881	35 588	3 203	38 437	3 459	35 309	3 178	30 316	2 728
Educational	9.0%	13 839	1 246	16 691	1 502	18 624	1 676	16 851	1 517	13 356	1 202
Religious	9.0%	7 740	697	7 522	677	7 197	648	6 177	556	5 156	464
Manufacturing	9.0%	32 264	2 904	40 215	3 619	52 754	4 748	56 296	5 067	37 478	3 373
All Other Private	**3.0%**	75 574	2 267	102 862	3 086	118 037	3 541	114 631	3 439	102 708	3 081
Public safety	3.0%	419	13	595	18	623	19	471	14	207	6
Amusement and recreation	3.0%	9 326	280	10 193	306	10 508	315	8 402	252	6 541	196
Transportation	3.0%	8 654	260	9 009	270	9 934	298	9 056	272	9 912	297
Communication	3.0%	22 187	666	27 488	825	26 343	790	19 712	591	18 226	547
Power	3.0%	33 654	1 010	54 115	1 623	69 242	2 077	76 064	2 282	66 601	1 998
Sewage and waste disposal	3.0%	305	9	408	12	665	20	468	14	421	13
Water supply	3.0%	477	14	516	15	466	14	319	10	664	20
Other	3.0%	552	17	538	16	256	8	139	4	136	4
Total Public Construction	**4.0%**	255 385	10 215	289 073	11 563	308 738	12 350	314 895	12 596	303 024	12 121
TOTAL		1 167 221	47 854	1 152 351	51 025	1 067 564	50 795	903 201	42 681	803 621	35 490

Source: U.S. Census Bureau. 2011. "Construction Spending." <http://www.census.gov/const/www/c30index.html>

47

5 Data Needs and Concluding Remarks

Direct Losses: The NFIRS data provided by the U.S. Fire Administration and the NFPA survey data both contain loss estimates made by fire response personnel. These estimates largely exclude the damage caused by wildfires beyond municipal jurisdictions; therefore, there is a need to quantify the value of these damages. There is also a need for damage estimates provided to NFIRS and the NFPA to be validated since they are not necessarily made by those with the expertise to assess the value of fire damages.

Indirect Losses: Indirect losses, including the temporary housing, missed work, and lost business due to fire, are difficult to quantify. There is not a comprehensive data source for indirect losses; however, estimates have been made on these types of losses. Indirect loss estimates for home fires has been studied by Munson and Ohls (1980) and Hall (2011) while Meade (1991a) and Hall (2011) provide some analysis of loss due to business interruption.

Other: There are a number of other costs related to the implementation of fire protection, as seen in Table 5.1, that are needed. Meade provides estimates for displacement costs, business interruption, and product liability/litigation insurance; however, periodic data on these elements are not collected. As seen in Table 5.1, a number of costs and losses have limited data coverage. For example, there are programs that manage fuels to reduce the risk of wildfire and there are a number of campaigns and education programs to prevent the ignition of fires; especially since a large percent of fires are human caused. Other limited categories include emergency responder costs, transportation degradation, aesthetic costs, environmental losses, and wildlife losses among other things. Some costs and losses are bundled with other categories. For example, the WPI estimate of the installation of fire protection in constructed facilities is estimated using the value of construction put in place, which includes additions, alterations, and reconstruction; therefore, it includes retrofitting as well as new construction. For this reason, these two categories are labeled as "estimated (bundled)." Also, the costs of fire personnel and training are bundled with all fire department costs. The costs and losses labeled as "limited/none" are good candidates for future research as are those labeled "partial." Those costs and losses for which direct data is available are labeled "yes" for data coverage. There is a need for many of the estimated values to be validated as many are estimated via dialog with industry experts. Although experts are a convenient source of information and are better than no estimate, these types of estimates are likely to have large errors, which remain unquantified.

Concluding Remarks: This report identifies significant data sources on the national fire problem, including data on fire occurrence and the associated costs and losses. With scarce and limited resources, it is important to optimize expenditures on fire prevention and response. In order to achieve this goal, it is necessary to understand the extent of fire occurrence, the costs and losses related to fire, and the associated stakeholders. That is, it is important for decision makers and land managers, who invest in mitigation techniques and strategies to minimize the cost and losses of fire, to understand the extent of fire occurrence and the costs and losses that actually occur. As previously discussed, there are some gaps in the data related to fire. This report identifies the areas where data is and is not available so that future research can focus on filling data needs.

Table 5.1: Cost and Loss Data Coverage

	Data Coverage	Cost/Loss Category
Costs	Partial Estimation	Fire protection research and design for constructed facilities
	Estimated (bundled)	Installation of fire protection in newly constructed facilities
	Estimated (bundled)	Retrofitting constructed facilities for fire protection
	Estimated	Maintenance and repair of facility fire protection
	Partial Estimation	Fire protection research and design for equipment and goods
	Estimated	Production of fire protection in equipment and goods
	Yes (bundled)	Fire service personnel and training
	Yes (bundled)	Fire service equipment
	Limited/None	Fire protection of structures against WUI fires
	Estimated	Evacuation services and planning
	Limited/None	Emergency responder personel and training
	Limited/None	Emergency responder equipment
	Limited/None	Wildland fuels management
	Estimated	Standards and codes
	Estimated	Insurance premium
	Limited/None	Fire prevention education/campaigns
	Limited/None	Aesthetic losses for fire protection design
	Estimated	Displacement costs (hotel and temporary shelter)
	Partial (bundled)	Fire code enforcement
	Partial (bundled)	Fire incident investigation
Losses	Partial	Direct property damage (net insurance)
	Estimated	Indirect property damage (net insurance)
	Estimated	Economic value of civilian fatalities
	Estimated	Economic value of civilian injuries
	Estimated	Economic value of fire fighter fatalities
	Estimated	Economic value of fire fighter injuries
	Limited/None	Environmental degradation of the atmosphere
	Limited/None	Degradation of the landscape
	Limited/None	Loss of recreational use
	Limited/None	Timber loss
	Estimated	Direct business interruption
	Limited/None	Indirect business interruption
	Limited/None	Deadweight losses
	Limited/None	Psychological losses (irreplaceable goods)
	Limited/None	Transportation degradation
	Limited/None	Infrastructure losses
	Limited/None	Utility losses
	Limited/None	Wildlife losses
	Limited/None	Income losses

References

1984 National Sample Survey of Unreported Residential Fires, Final Technical Report for Contract No. C-83-1239 to US Consumer Product Safety Commission, Princeton, NJ: Audits & Surveys – Government Research Division, June 13, 1985. Calculated from figures on pp. ii and v.

Apostolou, John J., David L. Bowers, Charles M. Sullivan III. 1978. "The Nation's Annual Expenditure for the Prevention and Control of Fire." Worcester Polytechnic Institute.

Federal Emergency Management Agency. 2010. *National Fire Incident Reporting System: Complete Reference Guide.*

Fire Family Plus v. 4.0. 2008. <http://www.firemodels.org/index.php/national-systems/firefamilyplus>

Hall, J.R. 2011. *The Total Cost of Fire in the United States*. NFPA. Quincy, MA.

Hall, Jr., J.R., U.S. 2010. *Unintentional Fire Death by State*. Report No. USS15, National Fire Protection Association, Quincy, MA, May 2010.

Hall, John, and Beatrice Harwood. 1989. "The National Estimates Approach to U.S. Fire Statistics." *Fire Technology,* May 1989. 13.

"History of the Dollar Value of a Volunteer Hour: 1980-2009." Independent Sector. <http://www.independentsector.org/volunteer_time>

Insurance Information Institute. 2011. *I.I.I. Insurance Fact Book 2011.*

Karter, Michael J. and Gary P. Stien. 2010. U.S. Fire Department Profile through 2009. NFPA. Quincy, MA.

Karter, Michael J. 2010. *Fire Loss in the United States During 2009*. National Fire Protection Association. <http://www.nfpa.org/assets/files/pdf/os.fireloss.pdf>

Meade, William P. 1991a. "A First Pass at Computing the Cost of Fire Safety in a Modern Society." NIST-GCR-91-592.

Meade, William P. 1991b. "A First Pass at Computing the Cost of Fire Safety in a Modern Society." *Fire Technology*. Vol 27, No. 4. 341-345.

Missoula Fire Sciences Laboratory. *Fire Behavior and Fire Danger Software.* <http://www.firemodels.org/index.php/firefamilyplus-software/firefamilyplus-downloads>

Munson, Michael J. and James C. Ohls, Indirect Costs of Residential Fires, FA-6, Federal Emergency Management Agency, Washington DC, April 1980.

National Fire and Aviation Management. 2011. "Web Applications." <https://fam.nwcg.gov/fam-web/>

National Fire Protection Association. 2011. "The U.S. Fire Problem." <http://www.nfpa.org>

National Interagency Fire Center. 2011. *209 Program: User's Guide*. <https://fam.nwcg.gov/fam-web/>

National Interagency Fire Center. 2011. National Fire and Aviation Management Web Applications (FAMWEB). SIT Reports. 2002-2010. <https://fam.nwcg.gov/fam-web/>

National Interagency Fire Center. 2011. "Historical Wildland Fire Summaries." 2003-2010. <http://www.nifc.gov/fireInfo/fireInfo_statistics.html>

U.S. Census Bureau. 2007. "Federal, State, and Local Governments." <http://www.census.gov/govs/>

U.S. Census Bureau. 2011 "Annual Survey of Manufactures." <http://www.census.gov/manufacturing/asm/index.html>

U.S. Census Bureau. 2011. "Construction Spending." <http://www.census.gov/const/www/c30index.html>

U.S. Census Bureau. 2011. "The 2011 Statistical Abstract: State & Local Government Finances & Employment: Revenue and Expenditures by Function." <http://www.census.gov/compendia/statab/>

U.S. Fire Administration National Fire Data Center. 2010. National Fire Incident Reporting System 5.0. FEMA, Washington, DC.